THE CITA

Scilly's own 'whisky galore' wreck

by
Richard Larn & David McBride

Published by
Shipwreck & Marine - Ropewalk House - Charlestown
St. Austell - Cornwall - PL25 3NN
Tel/Fax: 01726 - 73104

First paperback edition 1997
Second paperback edition 1998

© Richard Larn & David McBride 1997

All rights reserved. No part of this
publication may be reproduced, stored
in a retrieval system, or transmitted,
in any form or by any means, electronic,
mechanical, photocopying, recording or
otherwise, without the prior permission
of the authors.
The moral right of the authors has been asserted

Richard Larn & David McBride
Shipwreck & Marine
Ropewalk House
Charlestown, Cornwall
England PL25 3NN

ISBN No 0-9523971-1-0
A catalogue record for this book is available from the British Library

Cover illustration: The wreck of the m.v. Cita,
ashore on Newfoundland Point, St. Mary's,
Isles of Scilly, 29 March 1997, with the
Salvage Chief and ***Scavenger*** alongside.

Designed by Troutbeck Press

Produced and printed by Troutbeck Press,
a subsidiary of R. Booth(bookbinder) Ltd., Antron Hill, Mabe, Penryn,
Cornwall. TR10 9HH. Tel: 01326 373226. Fax: 01326 376061

CONTENTS

Introduction

1. The wreck of the m.v. *Cita*

2. Containers and oil

3. An island bonanza

4. The *Cita's* cargo and its salvage

5. The Salvage Chief, Forth Explorer & tugs

6. The *Cita* disappears

7. Legal aspects of the wreck

The wreck of the m.v. Cita on the morning of 27th March 1997, listing at 70 degrees to starboard, with her upper deck almost awash at high tide.

Acknowledgments: The authors thank the following for their assistance and co-operation in the preparation of this booklet. In the Isles of Scilly, Steve Watt, Tourism & Maritime Officer; John Nicholls, Branch Secretary RNLI. and pilot; Councillor Clive Mumford; Andrew Gibson, Environmental Officer; Mike Brown; and Gibson-Kyne for permission to use black & white photographs from the Gibson collection. Also, the Editor of Lloyd's List; Barbara Jones & Anne Cowne, Lloyd's Register of Shipping; David Thomas, Southampton Container Terminal; Bob Hurrell, Senior Watch Officer, H.M. Coastguard, Falmouth MRCC.; Sue Bradbury, PRO. RNAS. Culdrose, V. Robbins, Receiver of Wreck and Ivan & Dan Corbett, Tor Mark Press, St. Day, Cornwall. All colour photographs by David McBride(copyright).

INTRODUCTION

> 'Whilst we do not pray for shipwrecks, oh Lord, should a wreck occur, we ask that thou wilt guide it to the Isles of Scilly, for the benefit of its poor inhabitants.'

History tells us that the above prayer, or an adaptation, was regularly used in the Isles of Scilly. Life on the islands was, for centuries, one of deprivation and poverty. Their inhabitants fought a constant battle to survive, made endurable only by 'God-sends' from shipwrecks which infrequently brought them some of lifes necessities and just occasionally luxuries as well.

There have been around 900 recorded shipwrecks on or near the islands since 1300 AD, the best known being the fleet of Royal Navy men o'war led by the **Association** in 1707, and the tanker **Torrey Canyon** in 1967. The former was, no doubt, welcomed for the timber, cordage, canvas and barrels of food it brought the islanders, the latter most unwelcome, the resultant loss of the bulk of her 119,328 tons of crude oil into the sea being Britain's first and the world's worst oil disaster at the time.

The wreck still most talked about by islanders took place 87 years ago, though no one is old enough to have rememberd it. This was the Atlantic Transport Company liner **Minnehaha,** which stranded on Scilly Rock, north of Bryher, on 18 April 1910; of 13,443-tons gross, she carried 66 passengers, a general cargo and 243 steers. Unable to refloat her at high water, the salvors decided she must be lightened by unloading some or all of her cargo. With no adequate facilities by way of lighters, barges or undercover waterfront storage on the islands, the only practical solution was to order her cargo thrown overboard! Model ' T ' Ford motor cars, grand pianos, crated machinery, sewing machines, carpets, shop-tills, pencils, clothing and dozens of other items were dumped - the majority into the sea - but naturally a large quantity managed to find its way ashore. More than one Scillonian family still have a Singer sewing machine from the wreck in its original ornate wooden box, one certainly being in regular use until 1960. The **Minnehaha** remained aground for 23 days before being refloated and towed away for repair. There had never been such a bounty of general cargo from shipwreck. Almost everyone in the islands benefitted from it in some way, despite authoritarian Customs Officers, a Receiver of Wreck, police and Duchy Steward.

But that was almost a century ago, a distant fragment of island history. With an international decline in shipwrecks - the m.v. **Poleire** lost near Tresco in 1970 with an uninteresting cargo of zinc concentrate, being the last wreck of any tonnage here for 27 years - no islander in their wildest dreams could have imagined that the sea would bring them a shipwreck like the **Cita.** By comparison the **Minnehaha** incident pales into insignificance, for here was a late 20th century modern bulk carrier/container feeder vessel - literally dropped into the laps of the Scillonians whose cargo was not only 'general' but of high value, running into millions. The raw tobacco she carried was worth £3 million alone.

Most wrecks around the Isles of Scilly, like the ***Minnehaha,*** occurred on the off-islands or else in inaccessible or difficult locations, more usually amongst the infamous Western Rocks where fierce tides would sweep away into the Atlantic to be lost for ever any item not immediately recovered. But the ***Cita*** was different. She drove ashore conveniently at Porth Hellick, on St.Mary's, the most heavily populated of the islands, her containers drifting in at locations equally as accessible on St.Agnes, St.Martin's and other islands and locals of all ages took full advantage of this latest and unexpected 'God-send'. Not the first wreck in this location, the ***Cita*** was preceded by the s.s. ***Lady Charlotte***, a 3,593-tons gross coal carrying freighter, from Cardiff to Alexandria, wrecked only a matter of 400ft(120m) away on 14 May 1917 so that the two wrecks lie almost parallel with each other and, in time, parts of both ships will merge. Around a small headland to the west wrecked containers under Blue Carn have spilled Toyota vehicle parts on top of the s.s. ***Brodfield,*** another British vessel of 5,686-tons, wrecked on 13 November 1916. Like the ***Cita,*** she, too, was seriously off-course, on passage from Le Havre to Barry Docks. In her case the Captain had attempted to navigate by dead reckoning all the way from Start Point, Devon. But the ***Cita*** remains the wreck of the century - the millenium even, the like of which had never before been seen - and probably will never be seen again. The event, which commenced at 3.30am. on 26 March 1997, has already passed into island history. Already nothing remains to be seen but the ***Cita*** will never, never be forgotten.

As John Hicks, a well known Scillonian boatman said: "I have waited 43 years for this; something I thought I would never see. It's better than winning the lottery."

The Atlantic Transport Co's liner Minnehaha stranded on Scilly Rock, 18 April 1910.

*Wrecked in fog on May 11th 1917, the s.s. **Lady Charlotte** was lost between Newfoundland Point and the rocks under the Loaded Camel, the **Cita** sinking close to her port side.*

The stranding of the cruise-liner ss. Albatros, 16 May 1997.

Fifty-two days after the **Cita** went ashore, the 24,804-tonne gross Nassau registered cruise-liner ss. **Albatros** struck the North Bartholomew Ledge in St. Mary's Sound at 3pm. on Friday 16 May, whilst leaving the Isles of Scilly. Watched with interest by visitors and locals alike, she weighed anchor after a brief stay and steamed gracefully past Star Castle in the wake of the pilot boat, on passage to Cowes, Isle of Wight. She was then seen to lurch, stop, and settle down by the bow, having torn a 131ft (40m) hole in her starboard side. She limped back to the Roads and anchored, where an underwater inspection revealed worse-than-expected damage. For the second time in as many months members of the Coastguard Agency's Marine Pollution Unit flew to Scilly, whilst the islands Council Crisis Management team went to 'red-alert' in anticipation of a potentially catastrophic spill of 480-tonnes of heavy fuel oil within the islands. Fortunately the pollution was minimal, the oil being transferred to a tanker alongside. Over 500 elderly German passengers disembarked to Penzance on the **Scillonian III** and the **Albatros** sailed, with a salvage tug in attendance, for repairs at the A.& P. drydock, Southampton, which took several weeks. Why the liner strayed outside of the buoyed channel, and whether or not the accident could have been avoided had the local pilot been on the bridge, instead of leading the vessel in a pilot boat, is for the Marine Accident Investigation Unit to determine in due course following their inquiry.

One - The wreck of the Cita

The *Cita* was built in Germany in 1977 as the *John Wulff*. She changed name and flag in 1983, becoming the *Lagarfoss,* owned by the Laga Line Ltd, her managers the Iceland Steamship Company, registered at St. John's, Antigua and Barbuda, in the West Indies. In October 1996, Lloyd's Register of Shipping were informed that she had changed her name yet again, to the *Cita,* in the ownership of the Martin Shipping Company Ltd whose managers were Reederei Gerd. A. Gorke, of Vorderstrasse 11, Hollern-Twielenfleth, Germany.

Designated in shipping circles a bulk carrier motor vessel, she had a TEU (Twenty foot Equivalent Unit) capacity of 128, ie.she was certified to carry 128 x 20 foot long containers, or the equivalent in alternative sizes. She was built by J. J. Sietas Shipbuilding Co, Hamburg, was of 3,083-tonnes gross with an overall length of 90.53m (312ft), 14.53m (47ft) beam and depth in hold of 8.62m (28ft).

Her bow was ice strengthened since she had at one time traded in northern latitudes. The *Cita* had a single 3 cylinder, 4SA, 3,000 rhp.oil engine, reduction geared to a single shaft which carried a variable pitch steel propeller; she was also fitted with a bow thrust propeller. Her superstructure was aft, she had twin funnels, two decks and was fitted with the latest electronic navigational equipment.

The *Cita* sailed from Rotterdam to the Port of London on 21 March under Captain Jerzy Wojtkow, - a 44-year-old Pole born in Rzeszow - where she transhipped containers. She then left for Southampton, arriving on 24 March 1997 and sailing the following day. Weather conditions were fair, with winds in sea areas Plymouth and Sole force 3-4, north-west. In addition to the Master she carried a crew of seven, comprising a Mate, Chief Engineer, a Motor-Man, three Able Seamen and a Cook, all Polish. Her destination was Belfast where her entire consignment of 145 containers, holding general cargo, were to be unloaded.

On the Isles of Scilly fog closed in during the evening of 25 March. The forecast suggested little change for 24 hours which meant that there would be no flights in or out of the islands next day. The *Cita* reported by radio to her agents in Folkestone that all was well. The sea state was now force 6 with mist and drizzle. Nothing more was heard from her until Coastguard Officer Max Woods, on radio watch at the Falmouth Maritime Rescue Co-ordination Centre, received the following message at 3.35 a.m.on 26 March 1997:

"Vessel *Cita* aground in position Latitude 49.54.7N; Longitude 06.16.7W. have 8 persons on board, call sign V2QC. Taking in water, pumping out, fast on my port side and aground at the bow. 22 degree starboard list, am afraid of loss of stability. Request assistance."

The Maritime Rescue Centre immediately alerted John Nicholls, Branch Secretary of the St. Mary's lifeboat. Her crew were called out and under coxswain Barry Bennett she left her moorings just after 4 a.m. with a crew of eight men on board. At the same time a request for assistance was sent to the Plymouth RAF. Search and Rescue Headquarters who called out a Royal Navy (SAR) Sea King helicopter from RNAS Culdrose on the Lizard peninsula. Helicopter R193 (Lieut. pilot Pat Webster; Lieut.co-pilot John Duffy; Lieut.observer Gus Stretton; LAC/winchman Robert McKee, and POAC/diver Phillip Warrington) was airborne at 4.10 a.m, arriving on scene to find the container ship aground on Newfoundland Point.

The St. Mary's lifeboat *Robert Edgar*, an Arun class boat No 52-18, had already recovered six crew members when the helicopter commenced recovery of the one injured crewman by highline transfer. Captain Wojtkow, who had elected to remain with his ship, now requested by radio his immediate airlift off the wreck which was listing some 60-70 degrees to starboard. The injured man was therefore lowered to the lifeboat which meantime received some damage to her communication aerials from the overhanging hull of the wreck. The Captain was recovered by double-lift method from the port bridge wing then flown direct to St. Mary's airfield where the SAR helicopter shut down, awaiting further instructions. The *Robert Edgar* then returned to Hugh Town harbour, landing the survivors who were joined shortly after by their Captain. The men were accommodated temporarily at the Harbourside Hotel where they received food and hot drinks. They left the islands on RMS *Scillonian III* at 4.30 p.m.for Penzance. Prior to departure Captain Wojtkow, still wearing his slippers, requested that he be taken to the airport to see the wreck. On reaching the end of the runway overlooking Giant's Castle and Newfoundland Point, he stood to attention and saluted his ship with tears in his eyes. The owners of the Harbourside Hotel, Tony & Jackie Pritchard, commented 'They were very shaken when the lifeboat brought them in.'

Meanwhile, the injured man, Able Seaman Lech Koralewski, was taken to St. Mary's hospital where it was found he had a broken leg. The navy helicopter was then tasked by radio to medevac the casualty to Treliske Hospital, near Truro, but on arrival the fog was too dense to allow a safe landing. R193 therefore diverted to RAF St.Mawgan from where the casualty was taken by ambulance to Truro hospital. The other crew members remained in Penzance for three days, visiting their injured companion and making statements and depositions to representatives of the Marine Accident Investigation Unit and the ship's underwriters.

St Mary's lifeboat, RNLI. Robert Edgar (52-18) which under coxswain Barry Bennett saved seven men from the Cita.

Since the *Cita* went aground during the early hours of 26 March, it was not reported in Lloyd's List - the international daily paper reporting shipping and insurance news - until the 27th. when the following appeared in the Casualty Report section:

"London March 26 - Following received from Coastguard Falmouth MRCC: timed 0411, UTC: M bulk carrier *Cita* V2QC (3,083gt, built 1977), with eight persons on board, reported aground . . lifeboat is proceeding to the scene. Timed 0522, UTC: M bulk carrier ***Cita*** has come off the ground and appears to be sinking. All crew members safely evacuated. Timed 0611, UTC: M bulk carrier ***Cita***, Southampton for Belfast with cargo of about 200 containers: Vessel developed a 60-deg list around 0500, UTC. Some containers observed by local tug ***Pendragon B*** to float free. At 0600 hrs, vessel observed by ***Pendragon B*** to be listing 70-degrees and awash over two thirds of her length. Many containers floated free. Vessel remains aground."

Islanders and visitors collecting sodden clothing from a container in Pilchards Pool, Porth Cressa.

"Timed 1010, UTC: M bulk carier ***Cita*** with 30 tons of diesel and 60 tons of fuel oil on board, is hard aground on starboard side. Estimated between 80 and 100 containers are adrift in a south westerly direction. Inshore tug ***Scavenger*** is on scene and police and fire brigade are in attendance. A Coastguard emergency towing vessel is en route. One crew member has been hospitalised with a broken leg. Wind south-

The bridge and superstructure of the Cita as seen from seaward on 27 March 1997.

westerly force 5 (fresh breeze) visibility poor. Low water about 1300, UTC." Another report was timed 1124: "UTC: M bulk carrier *Cita* remains hard aground on Newfoundland Point, Saint Mary's(sic). Up to 100 containers lost overboard, 15 washed up on Saint Mary's, two washed up on Bartholomew Ledge and some drifting in Saint Mary's Sound. Securite navigation warnings being broadcast. Coastguard m.tug/supply vessel *Far Turbot* proceeding."

Next day Lloyd's List reported: "*Cita* (Antigua & Barbuda) London, Following received from Coastguard Falmouth MRCC, timed 0545, UTC: M bulk carrier *Cita* is still hard aground. M tug/supply vessel *Far Turbot* is now on scene. A meeting will be held this morning to discuss salvage plans."

The next report appeared on Monday 31st March:

"London, March 28: - the UK's Marine Pollution Control Unit had made removal of 90 tonnes of bunker fuel its first priority after the grounding of M bulk carrier *Cita,* off Newfoundland Point in the Isles of Scilly. Smit Tak, appointed salvor to the vessel, has chartered a local vessel *Pendragon B*, which took divers to the scene to inspect the *Cita* and seal any leaks. MV. salvage vessel *Salvage Chief* was making her way to the casualty and was expected to arrive tomorrow at noon. Pollution unit staff were on readiness on the islands to deal with any problems which might occur during the diving inspection. M tug/supply vessel *Far Turbot* and the Smit-Tak chartered tug *Falmouth Bay* were on the scene to recover floating containers. Falmouth Coastguard redirecting nearby ships."

The *Cita* with the vessel *Salvage Cheif* alongside, her divers carrying out an underwater investigation.

London, March 30 - "Following received from Coastguard Falmouth MRCC timed: 1210, BST: M bulk carrier *Cita* is still lying on her side, semi-submerged on rocks. Operations to remove her bunker oil are being carried out. There was a small, but insignificant, spillage yesterday. Work to clear up the spillage, using a boom, has been undertaken. Efforts to recover containers from the vessel are ongoing. The MPCU is still on the island."

On the 2nd and 3rd April, various reports included:

"Some of the cargo of containers have been salved by local fishermen . . Salvage company Smit-Tak B.V. working to contain wreck . . A number of floating containers, loaded and empty, are now at St. Mary's and off islands, Newlyn harbour and Falmouth. Receiver of Wreck being informed as much as possible. There is no clear information as to subsequent action regarding salvage and disposal at present. - Lloyd's Agents."

The only comment concerning the accident allegedly made by one of the crew whilst still on the Isles of Scilly said: "We all aboard were asleep at the time." However, a national newspaper reported that 45 year old Polish Able Seaman Jan Warciak was a little more forthcoming: "We were all asleep when she struck. There was a big shock. We look for damage and there was big hole amidships and very quickly coming in water."

The immediate concern of Scillonians and the authorities alike was of any possible oil spill, which could have a catastrophic effect not only on the local environment and wildlife, but also the tourist season, the Easter Bank Holiday weekend being only three days away. Thirty years earlier, on 18 March 1967, the islands had experienced the nightmare of a 61,236-tonne, fully-laden super-tanker named ***Torrey Canyon*** tearing its hull open on the Seven Stones Reef, discharging some 80,000-tonnes of crude oil into the sea only seven miles from St. Mary's. Whilst that spill heavily polluted every cove and beach in West Cornwall, by a stroke of luck the prevailing wind kept the oil away from Scilly, and the islands escaped virtually unscathed, but Scillonians live in fear of another such accident.

In Scotland, the inhabitants of Shetland suffered a similar experience on 5 January 1993, when another flag of convenience tanker, the 89,730-tonne ***Braer***, tore her hull open and broke up on Garth's Wick, near Sumburgh airport. She spilt 84,413-tonnes of Norwegian crude, 1,700-tonnes of ship's heavy fuel and 125-tonnes of diesel at the height of the worst gale in Scotland for 100 years. At Milford Haven, the tanker ***Sea Empress*** went aground in 1996, the cause of another disastrous UK.oil spill, whilst aboard the ***Exxon Valdez***, ***Amoco Cadiz***, ***American Trader***, ***Evoikos*** and the ***Aster*** were but a few similar tanker casualties.

On Monday 11 January 1993, the then Transport Secretary John MacGregor rose to his feet at 3.30pm. to make a statement to the House of Commons regarding the ***Braer*** disaster; and assured a crowded House, 'The Government strongly upholds the polluter-pays principle.' Thirty years after the ***Torrey Canyon's*** loss and five years after the ***Braer*** incident, there is still no international agreement regarding pollution - no wonder the Scillonians were concerned when the ***Cita*** hit the rocks of Newfoundland Point..

Two - Containers & Oil

The *Cita* allegedly carried 145 containers, many were 40ft long and nineteen were empty. Having assumed a heavy list to starboard from the moment she drove ashore at 3.30 a.m, many of the containers stacked three high above the upper deck and three below in the hold broke their securing lugs and fell into the sea, followed by others as the ship settled down on the rocks. By 9 a.m. the local salvage vessel ***Scavenger*** was alongside the wreck with the Captain and three crewmen of the *Cita* who went aboard to collect personal belongings from the wreck, the ship's log, manifest and other official papers. Soon other boats were on the scene having been alerted by the lifeboat crew which immediately took in tow floating containers and got them into Watermill Cove.

By 7a.m, with fog and rain prevailing and many islanders still unaware of the wreck at Porth Hellick, less than a dozen people stood on the high ground close to the rocks known as the 'Loaded Camel', looking down on the *Cita*, but by 9 a.m. the cliffs were crowded. One of the first on the scene was David McBride, a local diver, who commented: "The containers stretched in a line from Porth Hellick leading to Pednathise Head in the west; they drifted out and then came back in on the tide." Young and old alike stared in disbelief at the scene, the first shipwreck many of them had ever seen, whilst in the background was a dreadful cacophony, the banging and screeching of tortured metal as container after container drifted ashore on rocks already densely covered with hundreds of quarter-inch-thick, 8 x 4ft sheets of plywood. "I never thought I would ever again see such a sight in my lifetime as a ship driven in on the rocks in this way," said lifeboat Branch Secretary and pilot, John Nicholls. One container painted a vivid green and ashore near the cove, held one million plastic carrier bags destined for the Quinnsworth chain of Irish supermarkets. The doors of this container had already burst open leaving the foreshore awash in loose green and white shopping bags each of which, ironically, bore the message 'Help protect the environment.'

By 6 a.m. the entire hold of the *Cita* had flooded to sea level through an obviously large hole torn in her bow area caused by driving ashore at full speed. Whilst virtually sealed units, although their doors are not watertight, containers can hold a considerable volume of air and depending on their contents those with a buoyant cargo - such as packaged clothing, shoes, tobacco, handicrafts etc. floated out of the wreck into the open sea, remaining half awash. Others containing car engines, cast iron fittings, fork-lift trucks, tinned chestnuts, gravestones, batteries etc.either remained inside the hold due to their deadweight or else stayed afloat for only a relatively short period before sinking nearby.

Soon the scattering of sight-seers became a crowd as news of the wreck spread through the islands like wildfire, visitors and locals alike flocking to the scene. Hundreds of rolls of film were exposed and the well known island photographer Frank Gibson, whose family have a long history of photographing Isles of Scilly and west country shipwrecks, was soon in touch with the London press advising them

Sean Lewis, an islander, with one of the green and white Quinnsworth supermarket shopping bags which were still floating around the islands weeks after the wreck.

Islanders stacking up vehicle tyres, boxes of shorts and computer 'mice' before the next high tide washed them out to sea.

that he had early pictures. Unfortunately fog prevented any helicopters or fixed-wing aircraft from reaching St. Mary's that day so there was no immediate transport by which reporters and television crews could reach the islands. Fortunately, the RMS *Scillonian III* was in service and sailed at 4.30 p.m, taking off the mail, returning again that evening on an unscheduled trip chartered by the MPCU to bring over additional police from the Devon & Cornwall Constabulary, members of the MPCU with specialist equipment, Captain Keith Hart of Aquarius International Consultants Ltd, who represented the underwriters and various members of the press.

With memories of the *Torrey Canyon* disaster of 1967 still fresh in the islands, particularly with members of the Isles of Scilly Council and its Maritime Officer Steve Watt, there was very genuine concern.

This wreck had all the components of another oil pollution disaster, albeit on a smaller scale, bringing possible ecological and economic catastrophe to the islands and West Cornwall at the very beginning of a holiday season. Whilst not an oil tanker, initially it was uncertain what type or quantity of oil the vessel carried, and whether or not her cargo included dangerous or hazardous chemicals. Whilst Loyd's List stated in their report of 26 March that the *Cita* carried 30 tonnes of light diesel oil for her auxiliary machinery and 60 tonnes of heavy fuel oil, the Western Morning News of 1 April announced that the ship " - carried 90 tonnes of heavy bunkering oil as ballast, and it is the damage that this could cause to the islands' fragile environment that has caused the greatest fear." Little could be achieved until the arrival of the Antwerp-registered vessel *Salvage Chief* which was bringing with her commercial divers and the necessary equipment to pump out oil from the *Cita's* fuel tanks. But she would take at least another day to reach St. Mary's. In the meantime, Maritime Officer Steve Watt went on local and national television to express his grave concern for the islands' vulnerability. "We have been pressing, since the Donaldson Report in the wake of the *Braer* disaster, to get the islands protected status in terms of shipping routes. This once again shows the urgency of our requirement." Pressure had been growing for some time for a 25-mile exclusion zone to be set up around the scattering of tiny islands which lie in the middle of busy shipping lanes. Chairman of the Islands' Council, Mike Hicks, who is also a boatman, insisted: "We need protection by making sure these huge ships are kept well out at sea." Island councillor Clive Mumford commented: "It's another example of how vunerable we are."

Throughout the winter the emergency towing vessel *Far Minara* had been on

A police officer having a quiet word with one of the salvors.

charter to the Coastguard Agency's Marine Pollution Control Unit, to supply cover in the western approaches. For the past five months this versatile vessel, whose primary role is anchor handling in the offshore oil industry, would have been much closer to the islands but in the eyes of government, winter had ceased and she had been returned to her owners. The Agency, therefore, instructed her sister ship, the ***Far Turbot*** to proceed to the scene in case she was required to assist refloat the ***Cita*** whilst one of their aircraft overflew the area to establish at an early stage the degree of pollution and take photographs. But this operation was hindered by poor visibility. Unconfirmed reports from the pilot indicated a "finger of oily sheen in the sea," but by now oil was not the only concern. It was then revealed that seven containers held 18.5-tonnes of rechargeable batteries which pollution experts said: "Could cause a serious lead poisoning hazard to the local marine environment" which was yet another problem.

Floating containers full of potentially dangerous materials, butane-filled disposable cigarette lighters and industrial liquid plastic had been a target for the MCPU's spotter plane 20 miles south of Land's End only six weeks earlier. On 13 February the 58,000-tonne Hamburg-registered container ship ***Tokio Express*** lost 62 of her 2,800 containers overboard in heavy weather whilst en route from Rotterdam to Nova Scotia. Three of these carrying general cargo, which included beer, furniture and gardening tools, were fortunately towed into Falmouth Roads but the remainder were believed to have sunk. A report by the International Maritime Organisation estimated that of the 10 million containers around the globe now used to carry almost all international dry cargo, some 120,000 are supposedly bobbing around in the oceans of the world having been lost overboard.

Correspondence following the publication of this booklet brought a copy of the magazine 'Beachcombers' Alert' from Seattle, Washington, USA, which carries a report on the loss of the 5 million Lego toy pieces from the ***Tokio Express***, which would stretch ten miles.

Containers litter the shoreline, with valuable bales of raw tobacco lying amonst the rocks.

Visitors recovering Marks & Spencers dresses from two containers, whilst in the background the St. Mary's fire brigade wearing breathing apparatus prepare to open a container.

Three - An Island Bonanza

"People are going berserk," said Isles of Scilly Maritime Officer Steve Watt, referring to the *Cita's* cargo. "It is gradually disappearing in all directions. It is just like Whisky Galore." In fact it was not until around noon on the first day that people began to help themselves to what they traditionally saw as "God sends," reviving the old tradition of plundering a ship's cargo. At first there was a degree of uncertainty as to the legality of opening up the containers to see what they held, coupled with the fact that with the tide still high the containers were out of reach until low water, around 1 p.m. None of them were padlocked or chained, their securing bolts being fastened only with large plastic 'tie-wraps' into which had been hot-melted a serial number. Local fishing boats had already pulled two containers into Watermill Cove and when these were opened later, their contents were found to be boxed Ascot trainer shoes and rolls of pvc tubing. There followed a mad scramble, people gathering up as many shoes as they could carry, regardless of size. Most had lost their original packing having disintegrated when wet. Car boots, the backs of pick-up trucks and Land Rovers, trailers, vans, even wheelbarrows, were all piled high. One amusing aspect was the fact that many individuals had only odd left or right shoes or else sizes of no use to them. Groups of people bartering shoes could be found all over the island, with children predominantly trying on shoes to find pairs. Such shoes retail at between £36 and £48 a pair and for many hard-pressed families with children they were more than welcome.

Salvors making their way along the coast path from Rams Pit, carring boxes of shorts.

Eleven containers in total went ashore in Rams Pit or Porth Wreck and were attended by both the police and local fire brigade. The latter donning breathing apparatus, and cautiously opening one particular container, were confronted with huge white rectangular canvas bags fitted with zip-fasteners each weighing some 672lbs (305kg) holding raw tobacco leaf. It was the first of many such container loads, several of which burst open in the sea or on the rocks, spewing out hundreds of similar bales. It was part of a consignment alleged to be worth £3 million. With high water around 6.40a.m making it difficult, if not impossible, to get close to many of the containers on the rocks from landward, some hours elapsed before they were sufficiently

A smashed container at Peninnis Head, with bales of tobacco and wooden plate-racks, showing the enormous clean-up problem faced by the Isles of Scilly Council and MPCU.

exposed for people to investigate their contents and what an incredible Pandora's Box they found. Men's striped polo shirts and hooded sweatshirts; computer mice complete with cables; vehicle tyres; hardwood exterior doors, both glazed and plain; more plywood, plastic carrier bags; wooden toilet seats; clothing & ladies shorts!

The Isles of Scilly is still very much a old fashioned community, where people respect each other and property and leave their houses and cars unlocked. Hence crime, as such, is almost unknown. The only law enforcement officers are resident police Sergeant Russell Mogridge and one constable, two special constables and normally one Customs & Excise Officer and a Coastguard Sector Officer. By coincidence the Customs Officer had retired only the previous Friday and the only full-time Coastguard shortly before that so there was a dearth of officials to handle the situation, and those who were available were a little perplexed as to what to do. The police on site handed out copies of the Coastguard Agencies Wreck and Salvage report forms (TCA/ROW 1) and advised people generally of the law.

Many amusing stories emerged out of the situation. One special constable confronted a local councillor he had known for years standing guard over a large pile of car tyres he had collected from the rocks. "Name please, Councillor xxx," he said as if he were a stranger. Councillor xxx replied that he was only doing his bit for the beach clearance campaign - and continued adding to the pile! The regular constable, watching the frantic activity with some amusement was heard to comment: "If I catch anyone with bald tyres on their vehicle after next week, I intend to do them for laziness." A certain member of the Council received a telephone call asking what they were going to do with a container holding 1,500 wooden toilet seats. "I can't tell you as yet - the Council are sitting on that one," he replied!

Additional police sent over from the mainland spent most of their time making notes, taking the odd name of salvors and vehicle numbers, advising the public of the need to complete ROW 1 forms and send them to the Receiver of Wreck at Southampton. Another Councillor asked of a constable returning to the mainland: "How did it go, did many people complete your wreck forms?" The policeman replied: "Very well indeed, thank you. We have had a lot of forms completed and handed to us, but one thing I do find surprising, is the number of people here named

The authors in diving suits (David McBride left, Richard Larn right) holding some of the Ben Sherman shirts they salvaged from the seabed near Giant's Castle.

Robert Dorrien Smith!"(Robert Dorrien Smith of course owns Tresco Island on lease from the Duchy of Cornwall). One particular story, considered by the islanders as hilarious, concerned a local who in order to enter the half submerged container full of baby clothes and sweatshirts, took off his own shirt and laid it on a rock rather than get it wet. He then dived into the sea but on returning with his arms full of Marks & Spencer's baby-grows, found that a souvenir hunter had made off with his shirt thinking it was booty from the wreck! Several ladies went waist-deep into Pilchard Pool at low tide to recover clothing, shouting out remarks such as: "Who wants a large one?" Sweat shirt of course!

The "Cornishman" newspaper's island correspondent Clive Mumford, wrote in his piece headed "Stripping spree on a sunken cargo ship. Hopelessly outgunned, the tiny local police force attempted to take names and vehicle registration numbers of those salvaging." Often they would be given aliases. As chains of hands transferred cargo from foreshore and rocks to clifftop one could not help but see the police in the context of a store doorman giving directions. "Hardwear? Over there, sir . . Footwear? . . Watermill Cove . . Children's clothing . . Porthcressa . . Rubber goods? . . The Garrison."

Two containers full of tobacco drifted round St. Mary's, one going aground on the Garrison shore where it spilt it's contents on the rocks, the other ashore in the Roads, near the Newman, from where a group of local men towed it into Porthmellon using the **Lyonesse Lady** motor launch. After taking legal advice regarding salvage they put in a claim for the entire container which could have been very profitable had the tobacco not been burnt. Another container from the wreck drifted round to Pendrathen, St. Mary's and when opened was found to contain a large quantity of new, but empty, Calor gas cylinders. These were recovered and stacked ashore at Porthmellon, which had become a general receiving area for containers and cargo. Other containers drifted round to St. Martin's, Bryher and Horse Point, St. Agne's, with reports of others sunk at various locations such as Spanish Ledges, Bartholomew Ledge, Serica Rock and as far away as Rosevear in the Western Rocks.

Adults and children were now clambering over the rocks to reach the containers in a total "free-for-all", with islanders and visitors alike walking through the town carrying cartons and cardboard boxes loaded with ladies' shorts, childrens' clothing, shirts, training shoes and nightdresses, many bearing the famous St. Michael brand name of Marks & Spencer, others St. Bernard, its Irish counterpart, or Ben Sherman designer shirts, long and short sleeved, worth £40 apiece. One female visitor returned to the mainland on the **Scillonian** carrying a brand new car tyre in each hand. Others were flown out!

The Western Morning News reported one islander as saying: "It's a traditional pastime here but when the girls turn up at the pub tonight I wonder how many of them will be wearing identical dresses?" He went on to comment: "Youngsters worked their way through mountains of designer trainers to find a pair to fit, locals made off with scores of expensive teak doors and every car on the island is expected to have new tyres from the 4,000-odd found on the beach so far. A party mood prevailed more than 24 hours after the first of the containers washed ashore, and islanders and a large number of visitors, have enjoyed the greatest take-away bargain of all time.

Every clothes line in the Scilly's seems to have new shirts and shorts hanging out to dry. T-shirts, Actionman clothing kits and computer equipment were in abundance, including 25,000 computer mice. On Wednesday evening (26th) the five bars and pubs on St. Mary's were almost empty as regulars made the most of a smuggler's moon to return to the wreck in darkness to see what else they could find." Another observer stated:"Very few people appeared to be at work yesterday(26th) - building sites were abandoned, flower fields left, gardens empty, guest-houses left half painted . . . it was a good job the school is closed for the Easter holiday for if it was a normal day, the attendance would have been zero - and that includes the teachers!"

The wreck proved a major attraction, bringing considerable extra business to the Isles of Scilly, the Steamship Company, it's subsidiary Skybus service and British International Helicopters who laid on extra flights. Fortunately for all concerned the Western Morning News was running a promotion, offering reduced fares on the ***Scillonian***. By collecting vouchers one could visit the islands on a day-trip for £12.50. This brought in between 300 and 400 people four days a week, many just to see the wreck. There was also considerable demand to view the **Cita** from seaward and enterprising boatmen were soon running sight-seeing trips. However at the request of the Belgian Towage & Salvage Union's vessel **Salvage Chief** (725-tonnes gross) which arrived off Scilly during the night of the 28th, the St. Mary's Harbourmaster Jeff Penhaligan and Howard Wright of the MPCU signed a Duchy of Cornwall order declaring an immediate one mile exclusion zone around the wreck. This was to allow her divers to ascertain the **Cita's** condition and investigate how best to remove the oil still on board. It brought an immediate end to the boat trips.

Many parents with young families, struggling on wages far below the so-called national average, were delighted with an opportunity of free clothing for their children, sufficient for several years growth - if not generations. An unnamed father in his twenties commented to a Western Morning News reporter: "The way I look at it, if it was an oil tanker we would have a disaster on our hands, so why shouldn't we benefit from this? There were around 25 people on the beach when I arrived yesterday morning, young and old, many of them visitors to the island. A lot of the containers were open so the goods were already damaged by the salt and sea. I don't see how they could have been re-sold, but they were good quality makes. It was all good-humoured, nobody was fighting over things." He continued: "I heard reports of people passing goods down a chain of ten or twelve into cars. Yesterday, islanders were loading up with stuff, taking it into town and then coming back for more. They were taking trainers, clothes and car tyres by the car load . . . The police have been making an appearance now and again. They have warned people that the Receivers have arrived on the island and if they find out who has taken stuff they may search people's homes for it later. I suppose we might have to give things back, but then the police haven't stopped anyone from taking anything, apart from the tobacco . . . I missed out on Wednesday because I was at work all day, and when I got down to the beach all that was left was one damaged computer mouse, a vast amount of plastic wrappings and sodden cardboard containers. I know people have actually opened containers that were sealed but I wouldn't do that . . . Some of the goods in those crates could have been resold. I'd say about half the people down on the beach were visitors or holidaymakers. They have probably had a better time here because of it."

Report of Wreck and Salvage forms TCA/ROW 1, which salvors should complete and forward to the Receiver of Wreck at Southampton.

Four - The Cita's cargo and its salvage

There is, of course, nothing new about a ship going ashore and losing it's cargo to the sea; neither is there anything new about people's attitude towards wrecks, generally seen as "something for nothing." Whilst Cornishmen and particularly Scillonians have been portrayed in history as possibly the worst of the so called "wrecking" fraternity it is a well established fact that "wrecking" was equally prevalent around the whole of the British Isles and that the people of Kent, Yorkshire or Pembroke were no different from their west country counterparts.

When the *Cita* went ashore - and for several hours after - those in authority could only speculate on her cargo until such time as either the Master, ship's agent or underwriters produced her manifest.

From the moment the first container was opened near Porth Hellick and people saw for themselves the variety of goods carried it was obvious to all that here was an extremely valuable collection of manufactured goods. The ship was ashore for several days before copies of her manifest filtered down to the general public; even then four pages were omitted, presumably because the underwriters declined to reveal everything she carried, or the missing pages detailed potential environmental or physically hazardous material.

The following is a copy of the bulk of the Cita's manifest; some pages of which were never made available, not even to the Island's Council.

Bales of tobacco on Porth Mellon beach, with the barge Pine Light in the background loaded and ready to be towed to Plymouth. The tobacco was so contaminated with salt water it was later burnt.

nb. Ctns = cartons Pkgs = packages Ptls = pallets Pcs = pieces ★ = Not Recovered
Container No   Size   Weight kg   Commodity

Attached Rider for: Mitsui. Cita 24.03.97. To: Dublin
Container No	Size	Weight kg		Commodity	
MOLU2237595	20	4784	STC 156	Ctns shirts	
MOLU2492693	20	7180	STC 500	Ctns computer accs.	
MOLU2913175	20	2260	STC 185	Pkgs body boards	
TR1U1642414	20	4890	STC 518	Ctns lens blanks	★
MOLU2790928	20	2733	STC 156	Ctns body boards	
MOLU8051355	40	22047	STC 785	Pcs cylinders	
MOLU8175157	40	12320	STC 369	Pcs cylinders	

Rider for: Nedlloyd Cita 24.03.97. To: Dublin
Container No	Size	Weight kg		Commodity	
INBU3044969	20	16800	STC 14C/S	Cast iron products	
INBU3181719	20	16800	STC 14C/S	Cast iron products	
KNLU3181177	20	5071	STC 725	Pkgs tyres	
TPHU5193021	40	9846	STC 1280	Pkgs tyres	
KNLU4244310	40	10046	STC 1360	Pkgs tyres	★
OOLU5026685	40	9838	STC 1410	Pkgs tyres	★
KNLU4264954	40	10014	STC 1500	Pkgs tyres	★
INBU4816389	40	8468	STC 1056	Ctns garments	
KNLU4187729	40	7229	STC 886	Ctns garments	★

Rider for: Hapag Lloyd Cita 24.03.97. To: Dublin
Container No	Size	Weight kg		Commodity	
HLXU2027300	20	4740	STC 903	Ctns garments	
HLCU2660334	20	17104	STC 13	Plts plywood	
HLCU2249920	20	2000	STC 200	Ctns shoes	★
HLCU2253236	20	2071	STC 57	Ctns auto parts	
HLCU2602894	20	20000	STC 50	C/S granite	
OCLU4173311	20 OT	8390	STC 2	forklift trucks	
POCU4010333	20 OT	6140	STC 1	forklift truck	
HLCU4184094	40	4104	STC 117	Ctns auto parts	
HLCU4130992	40	22969	STC 2312	Ctns refuse sacks	

Rider for: P & O. Nedlloyd. Cita 24.03.97. To: Dublin
Container No	Size	Weight kg		Commodity	
OCLU1373893	40	13000	STC 439	Pkgs tyres	
OCLU0746380	20	3600	STC 268	Ctns dresses	★
OCLU0933231	20	7500	STC 166	Ctns toys & sundries	★
OCLU1448267	40	11200	STC 813	Ctns garments	
OCLU7026818	40 HC	16000	STC 1370	Pcs tyres	★
OCLU7028338	40 HC	16000	STC 1514	Ctns tyres	★
OCLU7033437	40	16000	STC 1472	Pcs tyres	★
POCU0615157	20	7400	STC 985	Ctns bathroom acces.	★
POCU7003066	40 HC	16000	STC 1900	Pkgs tyres	★
POCU7024752	40 HC	17000	STC 1492	Pkgs tyres	

POCU7050134	40 HC	16000	STC 1606	Pkgs tyres	★
POCU7060317	40 HC	17000	STC 1940	Pkgs tyres	★
POCU7077290	40 HC	17000	STC 2026	Pkgs tyres	
POCU0168758	20	4500	STC 318	Ctns trophy parts	
SECS9306774	20	2000	STC	Empty tank	

Rider for: NYK. Cita 24.03.97. To: Belfast

NYKU2125665	20	8680	STC 467	Ctns motor capstan	
TRIU29907049	20	18428	STC 14	Plts flooring	
TRIU3990717	20	18403	STC 14	Plts flooring	
NYKU2423783	20	4289	STC 174	Ctns VCR. parts	★
NYKU6658686	40	12277	STC 26	Plts satellite recr. parts	★
NYKU6016451	40	11382	STC 431	Ctns VCR. parts	
NYKU6983718	40 HC	16900	STC 767	Ctns VCR. parts	

Rider for: Sealand. Cita 24.03.97. To: Belfast

OSTU6556902	40	21000	STC 74	Pkts tobacco	
SEAU8088183	40	21000	STC 90	Pkts tobacco	
SPLU0419812	40	21000	STC 90	Pkts tobacco	★
GSTU7240230	40	21000	STC 90	Pkts tobacco	
INBU4640751	40	21000	STC 90	Pkts tobacco	
SEAU8132922	40	21000	STC 90	Pkts tobacco	
SEAU8220258	40	21000	STC 90	Pkts tobacco	

Rider for: P & O. Cita 24.03.97. To: Belfast

POCU0001694	20	18000	STC 28	Plts batteries	★
POCU0198602	20	15000	STC 285	Trusses jute yarn	★
POCU0345215	20	15000	STC 285	Trusses jute yarn	★
POCU0396613	20	13300	STC 177	Rolls fabric	★
POCU6702536	20	5400	STC 2400	Sets clothing	
OCLU0807193	20	8500	STC 16	Plts forklift truck parts	
OCLU0897160	20	6700	STC 198	Pkgs tyres	★
POCU0377975	20	12800	STC 29	Plts seat belt parts	★
POCU7700116	40	10400	STC 224	Ctns clothing	★
POCU7700292	40	8300	STC 1000	Pcs & 121 ctns garments	

Rider for: NYK. Cita 24.03.97. To: Dublin

NYKU7590628	20 RF	11511	STC 24	Plts diamiron temp units	
TRIU3998153	20	11191	STC 609	Pkgs power tools	★
NFLU2000031	40	13600	STC 16000	Ctns barbecue sets	★
NYKU6792563	40	15690	STC 41000	Ctns bbq. accessories	★
MLCU4074941	40	24468	STC 17	Plts polyester film	
MLCU4129366	40	24238	STC 17	Plts polyester film	
MLCU4311750	40	24238	STC 17	Plts polyester film	
MLCU4817667	40	24238	STC 17	Plts polyester film	

MLCU4851511	40	24238	STC 17	Plts polyester film

Rider for: Evergreen Cita 23.03.97. To: Dublin

EISU3249759	18732	28	Plts rechargeable batteries
ENCU2514460	18519	18	Plts rechargeable batteries
ENCU2822903	18519	18	Plts rechargeable batteries
GSTU3001921	10014	18	Plts rechargeable batteries
GSTU4997350	10014	18	Plts rechargeable batteries
UGMU9526368	18732	20	Plts rechargeable batteries

(please note that this cargo is hazardous)

EISU3326630	1500	270	Pieces panel and cases
PRBU2191001	5187	411	Ctns bathroom accessories
UGMU8477496	9300	810	Pkgs scales
EISU3297720	12437	585	Ctns inner tubes
EISU3334845	6300	199	Ctns handycrafts etc.
EMCU2997075	17200	956	Ctns plastic products
EMCU2641712	17313	1332	Ctns plastic products
UGHU8540000	20900	950	Ctns of canned water chestnuts
GSTU6166381	11000	448	Ctns terry towels
GSTU6216646	20100	20	Plts chemicals, harmful
SMOU2473545	20100	20	Plts chemicals, harmful
EISU1018299	16640	832	Ctns polyresin
EISE1083901	16309	60	Pkgs yarn
EMOU9816582	9000	700	Ctns shoes

Rider for: Geest Lines. Cita To: Dublin

19 containers   20 2000kg each   13 empty containers lost

The following are containers known to have been carried by the Cita, which do not appear on the manifest available, believed either unsalvaged or missing:

HLCU225479/9	20ft	8.2	Car engines
MAEU 738023/4	40ft	20.81	Drums Cab (possibly trapped in hold)
NOSU220726/2	20ft	2.0	Shirts
GSTU616638/1	40ft	11.0	Terry Towels (probably on Rosevear)
OOLU523906/3	40ft	19.5	Tobacco
MAEU792905/2	20ft	20.7	Wines

A closer examination of this manifest gives some idea of the value of the wrecks contents and its variety. For example, the most valuable aspect of her cargo was obviously 614 bales of tobacco, weighing 294 tonnes (294,000kg), said to be worth £3 million. The vehicle tyres, which included expensive Land Rover and tractor models totalled 195.5 tonnes (195,515kg), It is impossible to estimate how many tyres there were since they are described as being in packages & cartons but if we assume these represent individual tyres it is possible the *Cita* carried 18,596 in total. Clothing ranged from men's Ben Sherman shirts, through ladies' nightdresses, T-shirts, sweatshirts, ladies' and children's shorts, children's polo and T-shirts and dresses of all types, a total of 7,905 cartons and packages with an additional 900 cartons of ladies shoes. Rechargeable batteries, all of which were recovered, formed a considerable part of the cargo and weighed a total of 94.5 tonnes. With an incomplete manifest the only insight to the additional items carried are reports from salvors, both above and below water, of cargo seen or actually recovered.

Computer tower-unit cases, empty apart from their miniature loudspeakers and

Lightweight golfbags were found in one container near the wreck and with an active golf club on St. Mary's, were of great interest, but were eventually bought back by the importers and destroyed.

packets containing locks, keys, plastic CD.Rom panels and printed circuit motherboard guides were found spewed out of a smashed container in 30ft (10m) depth near Horse Point, St. Agnes. Close at hand were spiral wound leads for connecting computer keyboards and internal PC.cables. Twenty tonnes of tinned water chestnuts were on board and 1,500 wooden toilet seats were recovered as well as hardwood plain and glazed doors. In addition there were washing machine spares, wooden pepper-mills, hard-disc drive units, portable TV and satellite aerials, men's working coveralls and bibs, work jackets and trousers, 120 tonnes of pvc in 22km rolls and two containers full of hi-fi equipment.

Additionally there was 7.5 tons of toys and sundries. The toys included Actionman kits, with plastic ammunition boxes, camouflage clothing, miniature plastic boots and toy eagles. What sundries constituted is still unclear, but imported mug/coat racks, wooden dish drainers, fridge magnets and keyrings etc have been recovered in large quantities. The latter were intended for the Irish gift trade and include many thousands labelled "Irish thatched cottage," "Irish jaunting car," "Lucky Irish shamrock," "Lucky Irish leprechaun," "An old Irish blessing (May your cares be all behind you! May the luck of the Irish find you)". Another "Irish blessing" reads, "As you slide down the Banister of Life, may the splinters never point in the wrong direction." An interesting aspect of the key-ring fobs was that they carried a three-leaf clover card label reading "Irish souvenir brass keyring". But after only eight days underwater the tags were already heavily rust-stained by the "brass" and the printers, probably in Tiawan, had made a spelling mistake, labelling them as "brass kryrings!". One container, found in the Rams Pit area, held in excess of 300 top-quality, light-weight, Callaway Big-Bertha golf bags. With a nine-hole course on St. Mary's and a popular golf club these last items were very attractive, especially after it was established that their retail price was around £95.

The cylinders mentioned in the manifest, found in two 40ft containers, were empty but new Calor gas cylinders. Auto parts were listed, which comprised Toyota car wheel trims, disc brakes, brake drums, headlamps, side light mouldings, window-winding mechanisms, radiator thermostats, brake cylinders, wing mirrors, air filters and exhausts. The chemicals, described as harmful, of which there were 40 pallet loads,

The tobacco was contained in huge zip-up canvas bags, weighing a quarter of a tonne each.

Empty butane gas-cylinders at the Porth Mellon beach collection area, awaiting shipment to the mainland, with bales of tobacco and empty containers.

consisted of Golden Algicide and Leasure Time, Pool, Bright and Clear, for the treatment of swimming pools. To date one container holding a large number of new car engines, others with forklift trucks, (since recovered), refuse sacks, cast-iron products, motor capstans, power tools, jute yarn, fabric, seat belt parts, barbecue sets and accessories, bathroom fittings and Terry towels have still to be found since some 70 containers are still missing. Perhaps the greatest "prize" of all yet to be located by divers and raised - if it has not already gone to pieces - is the container known to hold 20.7-tonnes of French wine. Should this be recovered and the contents prove palatable (even 'near palatable' will do, according to some locals!) then the *Cita* could yet go down in Scilly history as it's "wine galore" wreck.

Throughout the summer and autumn months of 1997, anyone living on the Isles of Scilly who could dive was underwater, recovering cargo items from the sea bed or else searching for lost containers, never quite sure of what they might find. A great many containers, the empty ones particularly, simply floated off on the tide, as did some with buoyant contents, such as tyres, shoes, some clothing and possibly the wine. Other containers whose contents were of considerable weight, such as the cast-iron products, fork-lift trucks and granite, sank very quickly, and in general were found close to the wreck.

The granite was found by Keith Denby and Paul Gomersall in May, just off the port side of the wreck of the *Lady Charlotte*. It was later established that the granite had been quarried in South America, shipped on to India where it was cut, shaped and polished, then sold to a firm of importers in Belfast for resale to undertakers. There were 52 headstones, of a size acceptable for Catholic burials, plus grave surrounds, in a variety of colours including rose, black, pink and green. The importers have offered the salvors £8,000 for the intact items, but as yet no deal has been struck. Broken chunks of granite are currently being cut, polished and engraved on the mainland to take a clock or barometer, then offered for sale as souveniers of the *Cita* wreck..

What was thought to be bottled water on the manifest turned out to be tinned water-chestnuts, which had 'gassed' and spoiled in their tins underwater. The fork-lift trucks were recovered at an early stage, and with prompt attention were got running and by arrangement with the underwriters were sold on to someone in Hayle, Cornwall. Cast-iron products proved to be heavy duty flywheels and gears, which were sold back to Flextech, the importers, which after shot-blasting were sold on. Cloth mentioned in the manifest turned out to be a type of twill, for the manufacture of clothing for Marks & Spencer, but the cost to the salvors of having it professionally cleaned before it was suitable for re-sale was £1,500 per roll, and consequently remains on the seabed, as do the rolls of yarn. Armoured, underground electric 3 and 4-core cable filled one container. In 500m rolls, the exporters want it destroyed, the salvors claim it is re-useable, so it remains on St. Mary's, incurring storage costs. The infamous golf-bags, the subject of much speculation as to their worth (they sell for around £100 new), were bought back by John Swan's of Belfast, and destroyed.

Five - The Salvage Chief, Forth Explorer & Tugs

Informed by the Maritime Rescue Centre, Falmouth, of the wreck, the Coastguard Agency's Marine Pollution Control Unit based at Southampton swung into action. Whilst the *Cita* was not a tanker memories of the *Sea Empress* and its disasterous oil spill off Milford Haven were fresh in peoples minds. The MPCU immediately contracted with the Dutch company Smit Tak International to undertake salvage. They, in turn, hired the *Salvage Chief* from Union de Remorquage et de Sauvetage S.A. in Belgium, to go to Scilly and she was under way by 7.45 a.m. on 26 March.

A briefing took place in the Island's Council Chamber at 4.45p.m that day when a statement was rehearsed for Radio Cornwall due to be transmitted at 5p.m. In that broadcast, Steve Watt, the Isles of Scilly Maritime Officer, stated that by the time this vessel arrived on site if not delayed by strong winds in the Channel, the *Cita* would have been aground several days, a totally unacceptable situation. Two key members of the MPCU team arrived on the *Scillonian III* that evening, Howard Wright and Kevin Colcomb, the latter the Senior Scientist for the unit who stated that their priority was to get divers down and seal the wreck of oil leaks. Paul Glerum, a representative for Smit Tak International who had also arrived in the islands and seen the wreck, was of the opinion that salvage would not be possible in the short term and agreed the priority to be any oil leaks.

The Isles of Scilly Crisis Management Team, meeting in the Town Hall at

Two containers, still chained together, their doors smashed open by the sea, one full of packaged shorts, the other empty. The ease with which the sea ripped containers apart was frightening.

*Hugh Town Quay, with floating containers alongside and the **Forth Explorer** out in St. Mary's Roads.*

8.30a.m. on Friday 28 March, welcomed the news that the m.v. **Forth Explorer** had been put on standby by the MPCU to go to the Isles of Scilly. By chance this vessel was close at hand, on passage from Liverpool to Dundee, where she was to take up position as an oil-spill retrieval vessel in the North Sea. She is one of the most sophisticated oil spill recovery vessels in the world, carrying all the necessary equipment and technology on board to deal with a major pollution incident. The islands were extremely fortunate that she was not only in the vicinity but could be diverted so quickly. Had a serious oil spill situation arisen, this vessel was heaven sent to deal with it in the most efficient manner. Had it been available 30 years earlier when the **Torrey Canyon** spilt its contents then that situation would have been contained much earlier.

By 11a.m, it was reported that some half a tonne of oil had escaped from the wreck with a slick stretching from Newfoundland Point to Giant's Castle, near the end of the airport runway, and appeared to be drifting east. With the hired tug **Falmouth Bay** and fishing vessels criss-crossing the shipping lanes collecting loose-floating containers, the **Salvage Chief** - due to reach Scilly at 5 p.m - was tasked to go alongside the quay on arrival, pick up MCPU decontamination equipment and begin work on the wreck at first light, Sunday 30 March. The Isles of Scilly Council keep a modest amount of oil spill dispersant chemical and spraying equipment for such emergencies and with an estimated 4 tonnes of oil having escaped from the wreck two local boats, David Thompson's **Swan Dancer** and Andy Stephan's **K- Mar,** were put on standby. MAFF had given permission for spraying to commence if necessary. That same day more chemicals and spraying equipment in the form of two tracked beach vehicles arrived on the *Scillonian III* as a precaution, and everyone breathed a little easier.

The *Salvage Chief* put out early on Sunday, 30 March, attached lines to the wreck and commenced hot-tapping the vessel to extract first light, then heavy oil from her bunkers. The overall operation was expected to take 4 or 5 days. At a briefing in the Council Chamber at 8.25 a.m. Monday, 31 March, it was reported that 40-tonnes had been removed but that there were technical problems in replacing the oil in the tanks with water. A further meeting took place at 9 a.m. on Tuesday 1 April with Smit Tak International's representative who stated there was probably some 90 - 95 tonnes of heavy oil still in the vessel and that there was a need to move containers

One of the authors holds up some of the thousands of souvenir 'Lucky Irish' key-rings and 'fridge magnets found underwater near Horse Point, St. Agnes.

still in the hold at the next low water. That same morning the *Forth Explorer* arrived. In conjunction with a light aircraft her first task was to co-ordinate a survey for oil slicks. By Wednesday morning, 2 April, the salvors reported that the majority of the light oil had been extracted and that they now intended to put divers into the engine room that afternoon to evacuate heavy oil from the holding tanks. This could be difficult, they said, since the divers would first need to clear the crew's accommodation of floating bedding which was causing a serious problem, then get through to the engine room, locate and identify the tanks. With a small amount of oil having come ashore at Porth Hellick, a shore team commenced work at 7a.m. removing it from the beach manually as best they could. Men shovelled up contaminated weed

A whole container of these Ascot trainer shoes came ashore at Watermill Cove, whose contents 'vanished' in a matter of hours.

and sand into bags, trailors and dumper trucks.

Consideration was then given by MPCU officials to washing the contaminated beach sand, before returning it to the foreshore, a technique that required the use of a dedicated vehicle, rather like a huge cement mixer, fitted with a special rotating drum. Strangely enough, the MPCU had to seek approval from MAFF for permission to carry out such an operation since chemicals would be involved, and the authorities were cautious regarding the effects of dispersants used in this manner. The necessary vehicle was therefore hired by the MPCU, who agreed to pay the charges, and it set off from London, but unfortunately broke down part way to Cornwall, completing its journey on a low loader. The Council also requested additional manpower from the mainland to handle the clean up operation, and Cormac Ltd. stated they could supply up to 20 men. All this, of course, entailed considerable cost requiring the approval of the insurance agent representing the Isles of Scilly Council at each stage. The sum quickly escalated. At 6p.m, a further meeting took place in the Council Chamber to look at progress, at which it was reported that between 90 and 100 tonnes of oiled sand had now been collected from the beach at Porth Hellick, was ready for processing and that the situation appeared to be under control.

Fortunately, the weather forecast for the next few days suggested favourable conditions over the weekend, with westerly winds of 15 knots for Thursday, 3 April, changing to north westerly and increasing to 20 knots on Saturday, the 5th. then south easterly force 3 for Sunday, the 6th. which gave a two, possibly three-day window to complete the evacuation of oil from the wreck. The threat of possible leakage from the cargo's rechargeable batteries, still inside the *Cita,* caused the Council to

This bulk gas container was listed in the Cita's manifest as 'STC. Empty Tank', seen here on Porth Mellon Beach awaiting collection.

erect warning notices at Porth Hellick advising the public of hazardous substances. That evening, Thursday 3 April, Smit Tak International reported that two of the wreck's holding tanks were now empty, and that additionally their men had broken into the bow section and removed 440kg of ship's cellulose paint, carried on board for the vessels refit, which had been due immediately following her intended call at Belfast. The recovered oil was now being transferred into the **Forth Explorer's** holding tanks and this work in fact lasted three days. In the meantime arrangements were made for a 40-tonne crane to lift the mixer-vehicle from Penzance quay on board the **Scillonian III,** due to arrive at St.Mary's around noon on Friday, 4 April.

Porth Mellon became a collecting point for containers. These were towed in by the tugs **Pendragon B**, **Samos**, and Bryher's **Scavenger**, and then dragged ashore by a large JCB digger and pushed to the back of the beach. Gas cylinders from the wreck awaiting despatch to the mainland, joined them along with tobacco in bales, piled three high. Work on Old Town sea defences at St.Mary's by Samos Ltd, a westcountry marine construction company, had been in progress for several weeks using hundreds of tonnes of granite boulders brought over from Penzance on the deck of the steel barge **Pine Light** towed by the tugs **Pendragon B** and **Samos** . The collection of some 290-tonnes of tobacco, each bale weighing a quarter of a tonne, was a major problem for the underwriters. This was a very valuable and perishable cargo, which had to be warehoused as soon as possible. Samos Ltd, owned by 42-year-old Terry O'Sullivan, therefore contracted with Smit Tak International to transport salvaged cargo to Plymouth. By Saturday, the otherwise empty deck of **Pine Light**, a craft designed and built for the transportation of pine logs in Scottish waters, was stacked high with tobacco, 785 gas cylinders and 1,000 toilet seats ready to be towed to Victoria Wharf at Cattedown, Plymouth, once **Pendragon B** could be released from her recovery of containers.

Saturday 6 April saw all the local vessels capable of recovering containers from off the rocks or beaches working closely with the two salvage vessels, and a smooth operation was now in progress. After being dragged into the shallows, the containers were then towed out to the **Forth Explorer** who tied them of, until such time as they could be taken round to Porth Mellon. The **Scillonian III** arrived shortly after noon with the special mixer vehicle on board and the ten men hired from Cormac Ltd, who following a technical and safety briefing, were allocated areas of St. Mary's

Workmen collecting sheets of plywood and plastic shopping bags at Porth Hellick. In the background tugs are rounding-up floating containers.

foreshore to clean up, and set to work. By mid-afternoon the mixer vehicle was in use at Porth Hellick, washing oil-contaminated sand, pumping oily salt water into huge pvc.tanks. At the same time commercial diving groups from Scilly were out in perfect weather conditions searching for lost containers on the seabed and recovering cargo, mostly shirts, vehicle tyres and golf bags. A mainland sport-diving charter vessel, returning to Penzance with items of the **Cita's** cargo, received a visit from the police on arrival.

Following a good weekend, by Monday, 7 April, the weather had changed, the wind now east to south-east force 5, which increased steadily during the day to force 7- 8, forcing the **Salvage Chief** and **Forth Explorer** to seek shelter amongst the islands. The latter anchored in St. Mary's Roads, and used the opportunity to take on stores and offload containers and MPCU equipment. The forecast was that the weather would remain unsettled for the remainder of the week. At 2.27p.m. the Maritime Officer received a telephone call from William Thomas, Cornwall's Emergency Planning Officer on site to advise the Island's Council, who stated that he had he had seen between 1,000 and 1,500kgs (?) of polystyrene boards (possibly the body surfing boards mentioned in the manifest or else packing material) ashore near Giant's Castle, and asked if anyone with a camera could photograph the mess and fax the results to the newspapers. He felt a need to keep the media interest going as the insurers were by now assuming that things were under control and stable. Steve Watt therefore took his car and camera to Tolman cafe, Old Town, then walked the coast path, getting photographic evidence of the large amount of polystyrene ashore.

Considerable concern was now being expressed for the safety of the wreck. With an increasing south-easterly wind and rising sea state, the *Cita* was last seen at dusk working heavily on the rocks of Newfoundland Point and many feared she would not last the night intact.

One of two specialised beach-cleaning, all terrain, tracked vehicles sent to Scilly by the MPCU. seen here at Porth Hellick. Bags of oiled sand in front of the vehicle are awaiting a cleaning/de-oiling process in an adapted cement-mixer lorry also brought over from the mainland specially for this operation.

Computer tower units cascade down a rock slope underwater off Horse Point, St. Agnes

Six - The Cita Disappears

Following the *Cita's* stranding she was abandoned with half her length on the port side aground on rock leaving her stern section, including superstructure, accommodation and machinery, overhanging 130-145ft (40-45m) of water. At an early stage it was forecast that should the weather change it was likely she would break in two. The fact that so many containers floated free whilst others were removed by the *Salvage Chief's* crane, certainly prolonged the life of the ship by reducing her deadweight, hence allowing the hull to flex.

Whilst the future of the wreck was debated by the media, it was announced at a briefing in the Town Hall at 9a.m. on Saturday 5 April that as yet, no contract had been signed regarding the removal of the *Cita*, whether refloated and towed away, or cut up into sections and loaded onto a barge. Either way the cost would be enormous, requiring Smit Tak International to send for suitable barges and a heavy-lift floating crane, an operation which would take at least two months assuming the weather held fine that long.

Following an investigation of the site at 7a.m. on Tuesday, 8 April, these options became academic. The *Salvage Chief*, after leaving her anchorage off Tolls Island, reported that the wreck had all but disappeared overnight and that her divers would examine the wreck at low water, about noon, weather permitting, to ascertain what

*Deck of the **Salvage Chief** looking forward from her bridge, with divers in the water preparing to take oil suction pipes down to the **Cita**'s fuel tanks.*

had happened. At 4p.m. Smit Tak's representative said that only the *Cita's* port-side bow rail now showed above water. As yet they had no idea whether the stern section had broken off and disappeared to the bottom or was still attached to the bow, the sea being considered too rough to allow a diving team near the area. The MPCU followed up by saying they had found no surface sheen, no slicks, nor any obvious oil on Porth Hellick beach, and were very satisfied with progress. They were continuing to survey other beaches but with the salvors confident that 98% of the oil had been extracted any remaining was likely to be only light machine or diesel oil. This would normally break up and disperse quickly without chemical treatment.

Sand washing continued at Porth Hellick all day with a great deal of polystyrene and some oil still on the rocks below Giant's Castle, and large amounts of floating plastic reported at the western end of Old Town Bay as well as at the back of Rat Island. A quantity of plywood had also been recovered from Bar beach along with 12 bags of wreck debris. Fourteen days after the *Cita* had gone ashore, of her 145 containers, 19 of which had been empty, only about 80 could be accounted for. By now 53 were known to be on St. Mary's, four on the off-islands, another three reported by divers, plus a number towed to Penzance and Falmouth, the total still uncertain. One container was towed all the way to Falmouth by a fishing vessel, its crew probably looking forward to a decent salvage award. It was in the process of being lifted on to the dockside when the doors burst open, spewing out thousands of the now famous but worthless green and white "Real value - Quinnsworth re-usable" pvc.shopping bags - which all fell back into the sea!

It was never seriously considered that the wreck would be cut up and removed since the cost would far outweigh her value as scrap metal, even if her stern section could be rebuilt into a 'new' ship. Moreover the salvors would make nothing out of her cargo, now scattered far and wide. To simply abandon her where she lay was the easy option, but the Council's Chief Executive held that the islands must be returned to their 'pre-wreck- condition, concerned that not only would she become an eyesore as her steel slowly rusted, but a potential deathtrap to anyone tempted to board her. The strength of seas around the Isles of Scilly are such that no wreck is likely to remain in one piece for long. An interesting comparison can be drawn with the s.s *Lady Charlotte*, mentioned earlier. A steamship carrying coal from Cardiff to Alexandria, she went ashore in fog in almost exactly the same place on 14 May 1917. Records show that nothing of the *Lady Charlotte* showed above water by November that year, the sea having completely destroyed this 3,593 ton vessel in six months. Unlike the *Cita*, the *Lady Charlotte* drove ashore over her whole length, with no part overhanging deep water.

The stern section of the *Cita* now lies in deep water partly on its port side with 45ft(15m) depth clear over the bridge. There is a gap of some 100ft(30m) between her stern and forward section, the wreck having sheared just forward of her superstructure. Both sections have already moved and now lie almost parallel, very close to the remains of the *Lady Charlotte*. The stern section of the *Cita* will attract sport divers for as long as it remains intact, especially since it is in water deep enough to make it an interesting challenge, and offers an opportunity to enter the accommodation area to explore, or recover souvenirs.

A selection of cargo from the wreck of the **Cita**, *with a lifebuoy from her bridge, now on public display in the Charlestown Shipwreck and Heritage Centre, near St. Austell, Cornwall.*

For those not familiar with the Isles of Scilly, Porth Hellick has another historic shipwreck connection, since legend has it that the body of that famous Admiral, Sir Clowdisley Shovell, was found alive here on the foreshore on 23 October 1707 after the loss of his flagship **Association,** close to the Bishop Rock. A Scillonian woman on finding the body, then supposedly murdered him for his jewellery which included a particularly large and valuable emerald ring. She then buried him in a shallow beach grave. Presumably she mentioned to others that she had seen and buried the body of a gentleman since the corpse was exhumed and identified the following day. The Admiral's body was later transported to Plymouth on board HM man o' war **Salisbury,** then to his house in Soho, London, after which it was interred at Queen Anne's personal expense in Westminster Abbey. His widow, Lady Elizabeth, made enquiries in the islands concerning the missing ring, but to no avail, and the legend concludes that only when she was old and on her deathbed, that Mary Mumford confessed to a priest of her crime. It is said that the ring survives in Scilly to this day and that should it ever leave the islands a great disaster will befall them. At the back of Porth Hellick beach, close to where oil contaminated sand from the wreck of the **Cita** was processed, a simple monument stone marks the supposed burial site which in recent years has been fitted with a brass memorial plaque.

Seven - Legal aspects of the wreck

As with all shipwrecks, the *Cita* raised numerous legal and financial problems. The Isles of Scilly have a population of only some 2,000, hence its Council has limited financial resources with which to bear the cost of any clean-up operation, manpower, additional police, accommodation and other immediate expenses involved with a shipwreck. Whilst the MPCU. will pay most of the bills via the Government, who in turn have sued the owners, additional costs fell on the islands and protracted legal proceedings are inevitable before any settlement is reached with insurers and underwriters.

At the time of the stranding the cause was believed to be negligence, it being said that the bridge watch was asleep. This was officially confirmed at Southampton Magistrates Court on 13 October 1997, when both Captain Jerzy Wojtkow and Mate Jacekgawronski, having consented to summary trial in a prosecution brought by the Treasury Solicitor, pleaded guilty to an offence under Section 58 of the Merchant Shipping Act 1995 of doing an act, in breach of their duty, which was likely to cause the loss or destruction or serious damage to another ship or the death or injury to a person. The MSA's Prosecuting Solicitor interviewed both officers in the immediate aftermath of the incident, the Master admitting that a radar alarm designed to warn of any target within a pre-set range was not switched on. The Mate took over the watch at midnight from the Master and admitted falling asleep, being woken by the grounding. The Master was fined £2,000, ; the Mate £1,500, both officers paying an additional £250 towards costs.

The bare facts of the case were that this 3,083-tonne bulk-carrier had sailed due west at 14 knots for at least two and a half hours, with her steering on automatic pilot and the officer-of-the-watch fast asleep. He must have been asleep when the vessel passed Land's End, since shortly after he was required to change course due north, to enter the separation zone between the mainland and Seven Stones Reef. In those two and a half hours, the *Cita* cut right across two major shipping channels, and only luck prevented a serious mid-Channel collision. The crew, all Polish, were as follows:

> Captain Jerzy Wojtkow, aged 44 years
> Mate Jacekgawronski, aged 40 years
> Chief Engineer Marian Milaszewski, aged 40 years
> Motorman Jerzy Krajniak, aged 46 years
> Able Seaman Andrzej Stefanski, aged 47 years
> Able Seaman Lech Koralewski, aged 48 years
> Able Seaman Jan Warciak, aged 45 years
> Ship's Cook Cwiertnia, aged 29 years

nb: Should the reader question the sex of the Chief Engineer, he was male.

The major concern of the Isles of Scilly Council, residents and government agencies alike was of an oil spill or other pollution problems. Whilst the MPCU. has submitted a claim to the owners of the *Cita* for £192,285 in respect of costs incurred, the local Council had to bear the cost of container recovery, their shipment to the mainland, along with oiled weed and debris for disposal, the collection of floating cargo, and continues to assist in the cost of the totally unforseen environmental damage caused by the pvc.cargo Hence the Chief Executive took some pleasure in drawing the underwriters attention to the recently passed Merchant Shipping & Maritime Safety Bill that extends the ethos that the polluter pays to cover not just for oil, but other items of detrius and pollution such as containers and general debris. However the reality of the situation is somewhat different.

Under the innocent designation STC17 on the *Cita's* manifest, were five 40ft containers holding pallets of polyester film. Manufactured in Korea, the material, in rolls 3ft. wide and 11 miles in length, each weighing 400lbs, was destined to be magnetically coated, then turned into audio and video recording tapes. Each container held 68 rolls, a total of 340, whose overall length was in the order of 3,740 miles - sufficient to reach from Scilly to New York and part way back. Considered insufficiently valuable to warrant salvage, the pvc. was left on the seabed, and in time the containers broke up leaving the plastic sheet to roll around the seabed amongst the rocks. Slowly the sea commenced to shred the material, which then drifted weightless with the tide until it washed ashore. At each low tide it dried, and exposed to sunlight slowly became brittle, disintegrating into even smaller pieces, in the order of 2ins. square which were easily picked up by the wind and blown inshore. Countless billions of small fragments of plastic sheet littered the beaches, foreshore and fields in the vicinity of the wreck, and men had to be employed to pick it up, assisted by countless numbers of locals and holiday-makers. It has been estimated that some 5-600 hours of free labour has been given by these volunteers, in an attempt to stem the mass of plastic which was coming ashore on every tide. A salvage operation to recover the rolls of pvc still in the sea was mounted with financial help from the Council, the Duchy of Cornwall, English Nature, Countryside Commission and the RSPB, and at the end of 1997 a total of 218 rolls had been recovered intact, 40 damaged rolls, and assuming some 20 rolls went completely to pieces, that leaves an estimated 60 rolls or 660 miles on the bottom. Andrew Gibson, the Environmental Trust's Director, who has co-ordinated recovery of the pvc. said in December '97 that they were winning and that the volume of pvc. coming ashore had declined. 55 tonnes of pvc has since been sold and shipped to Dumfries for conversion into duvet and anorak lining material.

It must not be overlooked that the Isles of Scilly Council, under the Chairmanship of the Chief Executive, Philip Hygate, took a central role in the whole operation concerning the *Cita* wreck., some of his officers on duty up to 18 hours at a time. The Council Chamber was closed to all meetings, only authorised personnel allowed into the room, its walls covered in charts and maps, lists of strategic telephone contacts, tide-tables, tidal-flow charts, the ship's manifest and transport timetables. In addition there was a monitor showing video tapes from the MPCU. spotter 'plane, causing

The wreck of the Cita, as seen from the Loaded Camel at Porth Hellick, prior to her hull breaking in two just forward of the bridge.

Chairman of the Council to comment:"It looks like mission control!" As a result of this temporary 'exclusion zone' around the Council Chamber the Tourist Board meeting on 1 April was held in the Lyonnesse Bar of the Scillonian Club - the first time a Council meeting had been held in a pub!

Luckily, the Council had, 12 months earlier, initiated a two-day rehersal for such an emergency. Called "Operation Firefly" it established a scenario of a passenger ship on fire in St. Mary's Roads. The weather conditions were virtually identical to that of 26 March, with early morning fog preventing flying, the *Scillonian III* again proving to be the essential lifeline. The logistical planning and lessons learnt over those two days proved invaluable when it came to dealing with the *Cita.* The authority has its own Emergency Planning manual and this exercise tested and rehearsed the many aspects and agencies involved.

The Council deserved - and rightly received fulsome praise, not just from the local community and Councillors for the speed and efficiency of the operation, but also from the MPCU. and English Nature. Peter Hart, from Pembrokshire County Council who was involved in the Sea Empress incident, brought in to advise the authority on best practice in dealing not just with pollution difficulties, but also with the problem of pernicious insurers, told Steve Watt, the Maritime Officer, he wished someone like Peter Hygate had been in charge in Pembroke.

Regarding the recovery of such a rich and varied cargo, the media were quick to accuse the

Salvaged rolls of pvc. on St. Mary's, each weighing 400lbs and holding 11 miles of plastic film. The Environmental Trust successfully sought a purchaser in anticipation of recouping salvage costs.

actions of islanders and visitors as "looting" and berated those whose prompt action served only to prevent the cargo from total destruction by the sea. Had Scillonians done nothing and simply watched it all float out to sea, no doubt they would have been labelled lazy or lethargic - you can't win as far as the media are concerned! By implication the headlines of the Western Morning News on Friday 28 March, "Booty on the beach is worth a fortune" and the report that followed: " . . islanders, and increasingly high numbers of visitors, enjoyed the greatest take-away bargain of all times," suggested that this was theft on a massive scale. The police were extremely tolerant of the situation when in fact they could have been heavy handed and insisted that everything salvaged went to collecting points, searching vehicles and premises as they would have done 100 years ago. Expressing our personal opinion, had this wreck taken place on the mainland there would have been opportunists present in their thousands from as far away as London and Birmingham looking to make a profit. Then the police would have cordoned off the wreck and impounded everything. Because it took place in Scilly, with such a small population, where most individuals would be more likely to benefit physically than financially from the cargo, it was possibly accepted there was no need for an authoritarian approach.

The law regarding wreck and salvage is clear and simple, being enshrined in the Merchant Shipping Act. Everything in or on the sea has a rightful owner, and there is no argument that every single item from the wreck of the *Cita* belonged to someone - either the ship's owner, the original consignors of the cargo or whoever insured it, but certainly not the finder. The action of taking items from the wreck, from containers or found loose on the foreshore, floating or sunk, did not constitute an offence in itself, provided that the salvor did not then sell, use or otherwise dispose of the goods, destroy or take some other action so as to deprive the owner of what is rightfully his without consent. The Merchant Shipping Act 1995 requires salvors to submit details of recovered items to the Receiver of Wreck, based at the Coastguard Agency, Southampton, who has a duty to look after the interests of both the owner and the legitimate finder of wreck, within a period of one year following the salvors declaration. The options open to the legal owner are that they can advise the Receiver of their wish to either abandon or claim title to the goods. If they choose to abandon title then the item is deemed the property of the Crown, and taking into account its financial and historic value and the effort made in recovery, the Receiver of Wreck determines the outcome. In the latter case after proving ownership, the owner shall "on paying the salvage, fees and expenses due, be entitled to have the wreck or the proceeds of sale paid to him," MSA.1995 Chapter 21, Sect 239(1). Certainly, in law, it is not a case of "finders keepers", but in the case of the *Cita* where much of the recovered clothing for example was already being worn and had virtually no re-sale value, the underwriters did not seek recovery of their property.

The so called "looters" of the *Cita's* cargo could in fact be said to have saved goods which otherwise would have been lost to everyone. Those not familiar with the destructive power of the sea, should appreciate that three weeks after the wreck it was almost impossible to find anything in an exposed location that had not virtually been destroyed. Shirts were torn to shreds, tyres cut to ribbons, some even torn completely in two, whilst any commodity with a steel content was already heavily rusted. Even whole containers were torn open, as if made of paper, despite no real heavy weather or storms. One police spokesman said: "We are confident that the islanders have been displaying their usual helpfulness to the emergency services by removing the property for safe keeping, to avoid it being swept back into the sea at the turning of the tide," which was a masterful piece of diplomacy. The owners representatives later met with the police to make arrangements only for the safe storage of the bales of tobacco and gas cylinders.

Quoting one young father at Porth Hellick, who was picking up clothes for his baby daughter, he said:"A lot of the containers were open so that the goods were damaged by salt and the sea. I don't see how they could have been re-sold, but they were good quality makes," and of course he was right. No one was going to ask for the clothing back, then ship it to the mainland full of salt and sand, put it through a washing machine, then inspect, fold and repackage the clothing for sale.

It is said that some good comes out of every tragedy, and this was certainly true of the *Cita* wreck . A small group of islanders were horrified at the sheer waste of such vast quantities of childrens clothing, which incidently was still coming ashore at Porthcressa in December, nine months after the wreck. All the local mum's had more than sufficient for their own use and probably several generations to come, so a mass collection of the remaining garments was initiated, for donation to Romanian and other orphanages and charitable institutions.

Answering allegations of "looting" made by the media in their own particular way, they diverted attention from what was in truth a remarkable community effort. Sodden kiddies' garments were collected by the thousand and put into a one-and-a-half tonne skip placed on the beach at Porthcressa by local builder Richard Chiverton and his wife Jane, which was refilled no less than six times. From outside the Chiverton's home the clothing was then collected in wheelbarrows, shopping trollies, laundry baskets and bags, taken home by volunteers, then laundered and ironed before collection at a central point where it was bundled ready for shipment. It was said that:"There were lines of baby clothes drying all over the island," and the local launderette worked flat out washing salt-laden clothing at their expense for days. Notices appeared in shop windows seeking "washermen and washerwomen,", in a united and concerted effort to transform the image of shipwreck into relief for orphaned babies in Romania and Africa.Individuals who helped organise this concerted effort included Jenny Burrows, Alison Guy, the Chiverton's, Julia Ottery, the Methodist Minister, Rector Rev. Julian Ould and many, many others.

Following permission from the Receiver of Wreck to pass the clothing on, although no decision has as yet been notified between owners and salvors of clothing, 20 large packing cases were filled, four of which went to the Women's Refuge in Hallesdon, Birmingham, the remainder through the White Cross Mission, a Cornish-based organisation at Summercourt, run by the Rev. Patricia Robson. The Isles of Scilly Steamship Company generously transported the cases to Penzance free-of-charge, from where they were collected. Under the heading "Magnificent humanitarian effort" the local Guardian Newspaper reported that "The islanders of Scilly have responded in telling fashion." The Rev.Robson responded by saying:"The gift is very much appreciated because they are all new clothes. The best thing was that we also received at least 100 pairs of trainer shoes."

The need for some positive action by the government to prevent another *Cita* or a major oil-spill in the south-west - or for that matter anywhere around the British Isles - is still of paramount importance. There is no simple solution to what is now an international problem, best summarised overall as "safety at sea". Overall there would appear to be three easily identifiable areas of responsibility which have to be addressed, if any real progress is to be made on an international scale. The first and most important area concerns flag of convenience shipping, whereby vessels can be registered in countries such as Liberia and Panama, which allows shipping companies to lower their standards regarding the condition of ships, the number of crew carried, their pay, conditions and qualifications. The reason is of course, money. The second is

Countless scraps of polythene sheet mixed up with seaweed litter Port Hellick beach, which continue to come ashore although the volume is steadily declining.

constantly being unloaded onto the shipping industry, again to save money, and thirdly, the ship owners themselves, who interpret every attempt at improving safety, manning, standards, ship-construction, routing and electronic monitoring in terms of cost, which again is all down to money.

Most of the recommendations made by the Isles of Scilly Council are echoes of what was said following the **Torrey Canyon** spill, the **Braer** disaster in Shetland in 1993 and others. The Shetland islanders, in a petition demanded that the Government immediately underwrite the total cost (of major pollution problems) facing islanders and their council; enforcement of tanker exclusion zones; a full and open public inquiry after such incidents; unlimited liability for tanker owners; monitoring of tankers around the British coast; and 'an end to the whole flags of convenience scandal and the implementation of a global strategy for the safe management of tankers.' What did come out of the **Braer** disaster was Lord Donaldson's Inquiry 'Safer Ships, Cleaner Seas', to which the Isles of Scilly Council responded in 1997 particularly regarding marine pollution, identifying sensitive areas, pilotage and intervention. The outcome is awaited with interest; perhaps now there will be some real action by our government at last?

The overall problems are so many and diverse, that it is worth mentioning some which might otherwise escape the layman. The use of chemical dispersants on oil spills is all very well, if only we were certain of their effect on humans. As with organic phosphates, particularly sheep-dip, now recognised as harmful, in Shetland the MPCU used Dispolene 34S, which according to both the Norwegian pressure group Bellona and Green Peace had been replaced years before by Dispolene 36S, which was only half as poisonous. The former should never be used on rocky shores because it is so toxic. They also used 95 tons of Dispolene LTSW; 10 tons of BP Enersperse 1037 and 15 tons of Dasic 34S, whose chemical composition is secret - and so the public can only put its trust in the experts. Particles of dispersant chemicals sprayed on the sea will inevitably be carried by the wind to fall on land, affecting crops, animals and humans, but do we really understand the chemistry?

To quote from Jonathan Wills & Karen Warners book, 'Innocent Passage' (ISBN 1-85158-542-7) which should be read by every islander, the authors make a strong case for strategic radar monitoring stations around the British Isles. Rightly, they say: "The network of Coastguard lookout stations, manned in bad weather by knowledgeable local men, usually retired seafarers, has long since been dismantled. The Department of Transport in London decided that even those paltry, part-time wage bills were too high. The old lookout cabins become derelict, their timbers rotted in the winter gales. The local Auxiliary Coastguard volunteers are still in place, but their job is to help with rescues, not to stand 'bad weather watches' to guard their coast against rogue tankers . . the Coastguard Service got rid of many of its knowledgeable officers and replaced them with casual shift-workers - keen, committed, grossly underpaid people who, through no fault of their own, are less likely to have sea-going experience or detailed knowledge of the coast. The Coastguards have been reduced to a dwindling network of hi-tech radio shacks. In Shetland they do not even have a radar set." Similarly on Scilly, there is no Coastguard lookout or radar set, which if available could so easily monitor shipping around Land's End and in the Western Approaches. However, there are in fact only four civilian shore radars to cover the entire coastline of the United States, outside of harbour limits.

The positive identification of ships, even those whose watch-keepers were asleep or refused to answer radio calls would be simple, if compulsory transponder units were fitted, as they are on all commercial aircraft, who are not allowed to fly without them. The cost would be some £5,000 per vessel, a relatively small sum. The British Government is reluctant to go unilateral on anything regarding safety at sea, instead they prevaricate, set up a committee or an inquiry and refuse to do anything until the International Maritime Organisation agrees. From 1993 the US.Coastguard insist that every tanker entering US. waters is fitted with a 'towing package.' This is a system of heavy duty towing wires and bridles, fitted with shackles laid out permanently on deck, which can be 'snatched' easily by any tug, even if the tanker is un-manned or has lost power, but again it is all down to money. These towing kits are expensive. Since towing bollards have been known to tear off, being only welded to the deck, a steel 'kingpost' must be fitted, attached to the keel and rising up through every deck to forecastle level, which is an expensive modification.

In 1990 it was the US. who took unilateral action to phase in compulsory double-hulls for tankers carrying oil in bulk, operating in US. waters. The Oil Pollution Act of 1990 requires retrofitting or phasing out of all existing single hull tankers by 2015. Tankers are now being cut in two, new double hull forward tank sections being welded to the older stern sections carrying the machinery and accommodation, being the cheapest option.

Foreign crews on flag of convienience ships demand less pay than their British counterparts. At the same time ship owners reduce the number of men carried and impose unreasonable working hours on them. No wonder men fall asleep on watch. Such men often work a 100+ hour week, and as with the *Cita*, which carried only two officers qualified to keep a bridge watch, when there should be at least three, they never got more than four or five hours uninterrupted sleep and seldom more than six hours in 24, for weeks or months on end.

Lloyd's List recently reported under the banner headline 'World facing crisis on spill clean-up costs', that the day is fast approaching when the world will not be able to cope with the cost of cleaning up after major oil spills at sea. Unless more countries sign up to international conventions designed to protect member states with compensation, costs will spiral out of control. We live in a world totally dependant on oil. Without it transport, industry and the public utilities would grind to a halt. Perhaps now that the writing is on the wall, something will at last be done.

So many hardwood doors were salvaged on St. Mary's, they even found their way into local dinghies as bottom boards.

So many childrens garments were being washed on St. Marys, it was said that the mains supply dropped to 200 v AC!

Scillonian Jenny Burrows up to her waist in children's clothing in Pilchard's Pool, Porth Cressa, recovering garments for Romanian orphans.